MW01610786

Heroes of the Heart

Conversations with Veterans & Their Service Dogs

Heroes of the Heart

Conversations with Veterans & Their Service Dogs

Sally Hursey

Three Dogs Dancing

Published by:
Three Dogs Dancing Press
P O Box 693
Landrum, SC 29356
www.sallyhursey.com

In conjunction with:
Old Mountain Press, Inc.
85 John Allman Ln.
Sylva, NC 28779
www.OldMountainPress.com
Old Mountain Press eBook Division
www.oldmp.com/e-book

Copyright 2020© Sally Hursey
Cover design by Tim Davis
Edited by Carol Crawford
Cover photo of John Goughneour and Bruni by Joanna Goughneour
Interior text design by Tom Davis of Old Mountain Press
ISBN: 978-0-578-68012-5

Heroes of the Heart: Conversations with Veterans & Their Service Dogs.
All rights reserved. Except for brief excerpts used in reviews, no portion of this work may be reproduced or published without expressed written permission from the author or the author's agent.

First Edition
Printed and bound in the United States of America by Gasch Printing • www.gaschprinting.com • 301.362.0700
10 9 8 7 6 5 4 3 2 1

Heroes of the Heart
is dedicated to:

our veterans, their families, and their loyal service dogs
and the individuals, volunteers, and organizations
who support them

my hero and father, Col. John C. Williams, U.S. Army
who served in WWII

CONTENTS

Foreword

by
Ann Goodheart and Linda Williams
Service Dogs for Veterans Canine Assessment
and Adoption Team

HAVE YOU EVER SAID, "if only I had a magic wand, I would change this...or fix that?" Animal rescuers feel this way every single day because, in the world of shelters and rescues, doors open and dogs walk into an unknown future. Volunteers with these organizations share a love for animals, especially for those that are homeless. That is how Ann and I met. Having both worked in rescue for many years, we both knew that the large dogs are often difficult to adopt. We also share a love for the military, since both of our fathers served this country as pilots in WWII. The idea of using dogs to help veterans sounded like the perfect match for us. Forming Service Animal Project, our 501c3 organization, allowed us to pull shelter dogs for veterans suffering with Post Traumatic Stress Disorder (PTSD), Traumatic Brain Injury (TBI), and Military Sexual Trauma (MST). This became the answer we were looking for.

Three years later, "being in the right place at the right time" resulted in a chance meeting between Mary Ann Merrill, one of our Service Animal Project members, and Sally. One afternoon Sally was sitting in the reception area at her veterinarian's office waiting for her appointment. Sally overheard Mary Ann talking about our program. Interested in learning more, she contacted us and asked if we could meet and talk. The result of that meeting was the publication of *Molly to the Rescue,* a children's book Sally wrote that followed one of our rescue dogs from shelter life to graduating as an elite certified service dog.

This work has taken us on an amazing journey. Searching for dogs, we expanded our network of partners in North and South Carolina. We often heard, "So happy to see y'all! You want the BIG ones!" During our evaluations, we took the dogs to restaurants and stores, exposing them to a myriad of sights and sounds. Often this gave us the opportunity to explain to the public how a dog can make a difference in a veteran's life. We have seen the magic these dogs can bring. The veterans accomplish things such as returning to school, being able to go out in public, reconnecting with their families, restoring their confidence, and rejoining society. The dogs take these men and women down a path of healing to a better life.

In 2017, we became part of a service dog organization in Greenville, South Carolina, Service Dogs for Veterans, founded by Bill Brightman. Bill's organization trains each veteran and dog team with the goals of helping mitigate the effects of PTSD, TBI, and MST. They graduate as an American Disabilities Act (ADA) compliant certified service dog team. The organization serves Upstate South Carolina and Western North Carolina.

Over the years, we have enjoyed a wonderful friendship with Sally, and we value her continuing support of our work. When she reached out again regarding writing another book about our local veterans and their dogs, we were ALL IN! We hope, through Sally's interviews of the veterans and her heartfelt writing, you will have a better understanding of what these men and women have given to protect our freedoms. Each veteran's experience is different, so come along on their journeys through stories courageously shared.

Bill Brightman and Sabot

Bill Brightman,
Founder and Executive Director

I MET BILL at a party in 2017 when I was writing *Molly to the Rescue*, a children's book about a local rescue dog that becomes an elite service dog. Linda Williams and Ann Goodheart, formerly of Service Animal Project (SAP), were joining forces with Bill and his organization, Service Dogs for Veterans (SD4V), in upstate South Carolina to locate rescue dogs that could serve as possible companion and ADA compliant service dogs for veterans.

Bill is an unassuming, gentle man who joined the U.S. Navy during the Vietnam era and was trained as an Avionics Technician. He was assigned duty aboard the USS Independence and the USS Saratoga. At the conclusion of his four-year enlistment as an E-5, he earned a BSME degree on the GI Bill.

Bill spent twenty-seven years with a Fortune 500 company as a licensed professional engineer. He later founded two successful water purification businesses, and he became an instrument-rated private pilot. He credits Cesar Millan's popular television series, *The Dog Whisperer,* with creating his interest in working with dogs. He said he was captivated by Cesar's ability to diagnose and bring out the best in problem dogs while helping the owners understand how to meet the dog's needs. After watching every one of Cesar's shows, Bill was motivated to volunteer at the local Humane League in Pennsylvania to walk dogs to relieve kennel stress. The shelter offered training to volunteers with an experienced professional behaviorist to train on the rudimentary techniques to rehabilitate dogs that otherwise would not likely get adopted. "I absolutely loved it," Bill said. "The turn around stories for most of the dogs we helped were both amazing and heart-warming."

Bill said he was later offered a volunteer position on the adoption floor. "That was like icing on the cake because then I could work with kenneled dogs and interact with potential adopters." Little did he realize this set the stage for his next eye-opening experience.

Bill shared how he met two veterans on two different occasions who came to the shelter looking for a companion dog. He learned both veterans were struggling with adjustment to civilian life after their deployments. Both veterans admitted they thought a dog would fill the void from living alone and adjusting to their PTSD symptoms. "I knew at the moment of those two encounters where my interest in working with dogs was headed when I retired," he said. "I wanted to rescue dogs for my military brothers and sisters of this generation. I began my study of how dogs could be trained to help veterans with psychiatric disabilities."

When Bill and his wife Pam moved to Greenville, South Carolina, in 2013, he was supposed to be retired. At least that was what he had told his wife. He searched in the area for an organization where he could volunteer to work with dogs and veterans. He was amazed that there wasn't a program in the area helping veterans find a canine companion. He learned there were 38,000 veterans in Greenville County and 100,000 veterans in the five counties of the upstate. "I knew with this large number of veterans, and that according to VA statistics that twelve to twenty percent could have some degree of psychiatric impairment, there was an untapped market in need of companion and service dogs," he said.

The seed was planted. Bill began to ask who the "go-to trainer" was in Greenville County. The answer kept coming back—Connie Cleveland of Dog Trainers Workshop. "I quickly went to her website and got educated," he said. "Once we met and talked, she embraced the idea of having veterans join the evening group obedience classes at her Dog Trainers Workshop."

"What was it about Connie that clicked with you?" I asked.

"Connie's life has been dedicated to all phases of dog training, from puppies to future champions in sport and dog obedience to service dogs," he said. "She had developed her methods and high standards in her thirty years of experience. When I also saw she had a heart for including our veterans, I was ecstatic. She was a gift from God in helping launch Service Dogs for Veterans (SD4V) in 2014. She and her professional staff have been incredible supporters and friends over the years." Bill added that this relationship with Connie Cleveland enabled SD4V to launch and operate without the financial debt and overhead he would have had if he had leased or bought a facility for training.

"How has SD4V grown over these six years?" I asked.

"I started with two veterans in 2014, and now we average sixteen to twenty veterans and dogs in the four training phases. It has grown entirely by word of mouth," he said. "It's amazing, and I never intended for this to become a full-time job, but it certainly has, and I love every aspect of it!"

Bill smiled with pride as he told me about his program. "Our veterans come to us with VA service-connected ratings from fifty to one hundred percent permanent and total disability from PTSD, TBI, and/or MST. They are willing to work and not let their disability define their future," he said. "They graduate with pride after having gone through eight months of school with their dogs. These amazing veterans enjoy a depth of bond with their beloved battle buddies that defies description." Bill went on to add that the majority of the dogs are rescues, yet they graduate as tier one service dogs ready to perform three to six ADA compliant tasks.

"The transformation of the veteran-dog team from the beginning to graduation is nothing short of jaw-dropping," he said. "Veterans are now ready to take on life with confidence and security. Our entire SD4V team share in their pride."

The organization has grown every year, and in 2016, Bill realized he was old enough to be a grandfather to most of the veterans, so he didn't know if he would be able to continue to relate to the younger vets. Since then he has brought on younger folks to help with the day-to-day operations while keeping his role as the guide. He considers himself "the ass-kicking grandfather figure." He can look the veterans in the eye and tell them they will pull through. His age and wisdom help convince them it's true.

"The veterans really benefit and enjoy being around their brothers and sisters, so group training works well. After about the second or third session, the veterans start to relax, and the dogs desensitize to the other dogs," he said. "It's beautiful to watch the training evolve."

"Bill, don't you train companion dogs as well as service dogs for the veterans?" I asked. "How does that work?"

"Everyone takes the three months of beginner and intermediate obedience classes. Veterans who do not need the public access rights provided by the ADA for service dogs graduate after those two classes as companion dog teams," he explained. "Sometimes veterans start the program intending to complete the eight months of training, but life gets in the way, or their health becomes an issue, and they aren't able to complete the service dog training. However, they can graduate holding their heads high, having graduated as a companion dog team."

"What happens after companion dog training?" I asked.

"The veterans then join the five-month service dog training phase. This is where their dog learns a strict heel position and is transformed from a pet to a working dog under the ADA. This means the dog is individually taught tasks or actions to mitigate their handler's symptoms," he said. "These are taught in the advanced class, and after that the team moves into the public access training. In this last phase, the teams are supervised, and they demonstrate all they have learned in real life situations. The goal is to have them demonstrate and prove to themselves they can go anywhere and function as a team. This means that even though they are two entities, they function as one. Graduation follows this phase, where they receive their certificate, vest, and ID. Graduation is an enormous event, and the veterans are filled with pride and excitement at what lies ahead."

"In addition to heading up the charity, what other hats do you wear?" I asked.

"I enjoy teaching the advanced and public access classes," he said.

"Bill, since you don't have a building for SD4V, where do the training and meetings take place?" I asked.

"Our puppy, beginner, and intermediate classes are taught at Dog Trainers Workshop. We teach other intermediate and advanced level classes at Off Leash K9 facility in Greenville, South Carolina, that Chris Buol generously donates for our use," he said. "The public access classes are taught in various commercial establishments in the Greenville area. I work out of my home office, but when we have team meetings, our local Cabela's has been terrific about letting us use their conference room. These arrangements are by design and serve us well at this point. They enable our operating overhead to remain low so that most of every dollar donated goes to benefit our veterans and the dogs."

"You have thought of everything. What a difference you've made in these veterans' lives," I said.

"Thank you, Sally. I think so, too. Honestly, more than anything, it's humbling to me to be doing this important work, and I'm deeply grateful," he said. "For me, it boils down to being available and willing to serve, and being guided by my faith that God is leading me in this effort."

"Are there other things you would like folks to know about Service Dogs for Veterans?" I asked.

"First of all, we have eighteen wonderful volunteers. They are passionate about our veterans and their dogs, and they've made the organization what it is today. They connect with and encourage our students at every stage of the training. They are also a big reason for our expenditure efficiency of over ninety percent because they do much of the administrative work," he explained.

"The second thing I want to share is the phenomenal support we get from the Greenville community. Our incredible financial donors have fueled our ability to grow the program. They enable us to offer scholarships that cover ninety-eight percent of the program costs. The third point is we are a bit unusual in that we train veterans to train their dogs. It's hands-on from day one. This is very different from the traditional model where professionals do all of the training, and then pair the dog with the veteran. We love our model because the veteran is invested in their dog's success, and they are given the chance to accomplish something pretty amazing. A disabled veteran trains a rescued dog to be their ADA compliant service dog!"

Bill ended our interview by saying that when asked how dogs help our veterans, his short answer is "dogs touch the heart and engage the brain like nothing else." He added that most veterans who have trained with their dogs would say, "My dog is my medication without any side effects, except my undying heartfelt appreciation."

Ashley and Aubrey

THE SMELL OF FRESH COFFEE lured me into the quaint coffee house in Clemson, South Carolina. I ordered a nonfat latte and sat by the window to watch for Ashley and Aubrey. Clemson University students lounged around talking and enjoying the beautiful sunny morning. Summer semester always seems less hectic at a university. A table opened up on the patio, so I went outside to wait. After a few minutes, a petite girl with blond and purple hair came into view. Walking in step beside her was a large, apricot colored standard poodle. I immediately knew it was Ashley and Aubrey.

I introduced myself and got Ashley a coffee. Ashley had Aubrey settled in front of our table on the patio. "Where do you want to start?" she asked.

"How did you meet Bill and get involved with Service Dogs for Veterans?" I asked.

"I met Bill in 2016 while at Tri-County Tech before I came to Clemson. I had very little social life and I was isolated. I knew a service dog could help. I tried seven different dogs, and they didn't work out. I even tried my own adopted dog Riley, a boxer mix, but he was too hyper," she said. "Bill had Aubrey available because she had been gifted to Service Dogs for Veterans. An older woman who passed away had originally owned her. Then a family took her in as a pet, but it didn't work out. The father had a special place in his heart for Aubrey, but to keep peace, he donated her to Bill. When Bill said he had a female standard poodle, I said, 'I want her!' I drove Riley up to New York to live with my sister, and I drove back and got Aubrey."

"She seems so calm and well behaved," I said.

"As soon as I met her, I knew she was perfect. She sat down and leaned against me. She is bubbly, loving, and has a sassy attitude to match," she laughed. "She is a godsend in so many ways. Aubrey was already well trained. She had never been on a collar, but she is smart and learns quickly."

"What does Aubrey do to help with your symptoms?" I asked.

"She is trained in grounding techniques such as push, snuggle, up, and block, where she stands or sits in front of me. These are the trained tasks she has, but she helps in so many other ways," she said. "Aubrey offers constant support, and having a living creature to be responsible for motivates me to step out of bed on the days I had rather not."

"How has Aubrey changed your life?" I asked.

Aubrey leaned against Ashley's leg and gazed into her face. "My life has basically turned around completely since receiving and training with Aubrey. I was having regular panic attacks and avoiding going to class. With Aubrey, I have continued to go to Clemson and improved my GPA. Currently, I am pursuing two degrees, a BS in psychology and a BS in criminal justice. I am the president of the Clemson Student Veterans Association, and I volunteer with the Purple Heart Homes and with Tigers Together to Stop Suicide, a research and advocacy team."

"It sounds like life is good now with Aubrey by your side," I said.

"In September of 2018, I was awarded the opportunity to attend a leadership conference with Student Veterans of America in Washington, DC," Ashley said, beaming with pride. "In January of 2019, I also got to go to Orlando, Florida, with ten other student veterans to attend the Student Veterans of America National Conference."

"Did you take Aubrey with you to Orlando?" I asked.

"I decided not to take Aubrey into that chaos. Aubrey has given me the coping skills and the confidence to attend a conference with trusted friends; however, being away from her made me realize how much I count on her every day, and I missed her terribly. Aubrey has helped me step out of my shell."

"How many veterans are in the Student Veterans organization at Clemson?" I asked.

"There are fifty veterans in our organization and 350 veterans on campus," she explained. "We have our own place on campus, and Aubrey is kind of like our mascot. She loves to run around the Veteran's Center and greet people. We are a social and philanthropic organization and want to give veterans a place where they are understood."

"I am impressed," I said, leaning forward. "I hope more student veterans will get involved. Do any of the other veterans have service dogs?"

"Several other veterans have gone through the program. Cody, one of our vets, and his German shepherd started going to Service Dogs for Veterans, and now he is doing better. Seeing a service dog on campus is rare. Aubrey has had the opportunity to educate people about service dogs and the difference they can make in your life."

I noticed Aubrey panting from the afternoon sun. We moved to a table in the shade and gave Aubrey some water and then continued our conversation.

Suddenly, Ashley had a serious look on her face. "In December, Aubrey had a seizure at the Veteran's Center, and all the vets jumped in to help," she said. "In January and February, she had two more seizures, and in March she had three in one day. She was diagnosed with epilepsy and put on meds to control it. It made her groggy at first, but now she is used to it, and she's okay."

"I'm glad she is doing so well now. You would never know she has a problem," I said. "Let's switch gears and talk about your military service. How old were you when you enlisted?"

"I enlisted before graduating from high school at eighteen. I was in the Delayed Entry Program and went to basic training at nineteen," she said. "I was conflicted about what I wanted to do with my life, and I knew if I went to college, I would spend a year trying to figure out what I wanted to pursue and end up dropping out."

"Where did you serve?" I asked.

"I served in the U.S. Navy, and I was stationed in several locations. First of all, I was stationed in Charleston, South

Carolina. That's probably why I ended up back here. Then I went to Ballston Spa, New York, and finally San Diego, California. In California, I deployed on a nine-month assignment onboard the Aircraft Carrier, USS Carl Vinson, CVN-70. Our duty was to launch jets to provide air support for boots on the ground and perform recon missions. We supported Operation Enduring Freedom in the Persian Gulf."

"What role did you play on the ship?" I asked.

"I served as a Nuclear Machinist Mate with an additional specialization in chemical and radiological controls of the reactor, also known as an Engineering Laboratory Technician. This means I was in charge of all chemistry and radiation levels of three separate water systems to ensure longevity and proper operation and to minimize radiation exposure to the crew," she answered.

Ashley could see my surprise. "That sounds so complicated. I can't imagine being responsible for radiation levels onboard a ship."

"It was a huge adjustment getting used to being on a ship with 5,000 people. I was the only woman in my division of thirty sailors. While I was deployed, I was assaulted by someone I had worked with on the ship," she said, looking away.

"Ashley, I'm so sorry," I said. "How did it affect you?"

"Prior to my discharge, I was referred to an inpatient substance abuse rehab center. This thirty-five-day intense therapy changed my life," she said. "Located in the same building where I was receiving therapy was the OASIS program, an inpatient Post Traumatic Stress Disability (PTSD) clinic. Each class raised a PTSD service dog. While I was dealing with my own problems, the idea of a service dog kept popping into my head."

"Were you medically discharged from the Navy for PTSD after returning to the States in 2015?" I asked.

"Yes. When I returned to South Carolina, I knew one person, my friend Alan. I relied on him to go everywhere with me. Now I jokingly refer to him as my 'service human,' because I couldn't handle the anxiety of going on my own. I finally knew a service dog could help me."

"When did you decide to go back to school?" I asked.

"I decided it was time to continue my education in 2016 and enrolled at Tri-County Technical College before going to Clemson University. Now I have a double degree in psychology and criminal justice, and I plan to get a master's degree."

Ashley said her life had turned completely around since receiving Aubrey. With Aubrey by her side, her GPA had improved, and she had taken on leadership roles. She had also reduced her meds from five a day to one a day.

"It's been quite a journey in the last thirty-eight months. I'm not sorry for anything, because it's made me who I am," she said. "I wouldn't wish it on anyone, but I am okay now. I love Service Dogs for Veterans, and I can't say enough good things about them."

Ashley and Aubrey support Service Dogs for Veterans by attending their events such as Puppies and Patriots 5K and 1 Mile Dog Walk and speaking at organizations. She told me she would be speaking about service dogs at Clemson in the fall.

If you attend a Clemson game, be on the lookout for a large, apricot poodle sporting the Clemson colors of purple and orange. Her handler will be a petite blond with purple strands in her hair, walking with her head held high.

Aubrey

John and Bruni

I MET JOHN AND HIS BIG YELLOW LAB, Bruni, at a Spartanburg restaurant. We got settled at a table, and Bruni lay at John's feet. He was probably the largest service dog I had met. I knew John had enlisted in 1984 when he was only seventeen years old, and he had served as an Army Medic until 1991.

"What do you want to talk about?" he asked. "I don't know if you know it, but I was just terminated from my job."

His statement caught me by surprise. "I didn't know you lost your job. I am so sorry to hear that bad news. What happened?"

"It seemed like my job was going well, but things fell apart. There are benefits and challenges to having a service dog," he said. "My job took me all over the southeast, and Bruni traveled everywhere with me. My company had just hired a woman who was excited about educating everyone about PTSD, but she left, and the new woman they hired put the brakes on it. Bruni and I got left by the wayside. Two people in the office were allergic to dogs, so I got moved to an office by myself. Being isolated fed my PTSD, but at least I had Bruni with me. Things got worse, and they let me go."

"Let's talk about your time in the service," I suggested trying to change the topic.

"I was a troubled teenager from a divorced and broken home with no other options," he said. "I really had no intention of serving again once I got out in 1991, but 9/11 happened. I went back in 2005 at the age of thirty-nine but went into engineering instead of being a medic. The second time around, I wanted to make a difference rather than sitting on the sidelines. I'd become passionate about the mission to keep extremism off our shores."

"Are you still in the military?" I asked.

"I'm currently still in the Army Reserves, but I've gone as far as I can, and I'll be getting out soon. I've had a total of sixteen good years," he said.

"Where did you serve, and in what capacity?" I asked.

"I was in Ramadi and Fallujah in 2006-2007 with the 718th Engineers out of Fort Benning, Georgia, and tasked out as a surveyor to the Marines most of the time. Then I deployed to Afghanistan with the 415th Engineer Facilities Detachment in 2010-2011 in Sharana, a large forward operating base, in Paktika Province," he said. "We belonged to the 82nd and 101st during that deployment. I was in the cities of Iraq for 222 days and saw the worst of humanity."

"Where were you when 9/11 happened?" I asked.

"I was in college in the Pittsburg area majoring in production design, and I was single," he said.

"When did you realize you needed a service dog?" I asked.

"Later with a wife and kids, my family started to notice problems at home like screaming at the kids, getting frustrated in traffic, and sometimes almost passing out at the wheel. The straw that broke the camel's back was an incident in traffic when I lost it with the family in the car," he said. "I realized how much I had scared my wife and two young sons. My issues revolved around car travel, traffic, and sudden jolts of anxiety while driving. I also have difficulty in large uncontrolled crowds. I have to be in the back of the crowd. If there's chaos, I go into a complete meltdown. I suffer from depression and PTSD. I went through therapy for about one and a half years until my therapist left the VA, then I went about four years without therapy. She gave me a lot of tools to cope with, but a big chunk was still missing in my life."

"How did you get Bruni?" I asked as I gazed down under the table where Bruni was lying quietly.

"I called Service Dogs for Veterans, and Bill asked me lots of questions on the phone. I liked the idea of training with a dog. I thought this is the way to go," he said. "Then Bill came to the house to interview me and my wife. At first my wife was saying 'No more dogs,' because we had a dog before Bruni that was a nightmare. Bill convinced her to give it a try, and now she can't live without Bruni. Two weeks later when Bill brought him to the house, there was an instant connection. He is amazing; he is a dog from heaven. He never gives me any trouble."

"I bet your boys love him!" I said.

"My little three-year-old went nuts over him. He tried to ride him," he laughed.

"How does Bruni help you cope?" I asked.

"My boy Bruni is my constant companion and soul buddy. He rides with me when I travel, and he pops up from the back seat the minute I hit traffic, and so far, I haven't had an incident while driving," he said. "The minute his curious face nudges my arm, my blood pressure drops like a rock. He always keeps me grounded, and I'm able to take those deep breath moments that help me overcome the irrational fight or flight impulses. I have anger issues, but I am much calmer thanks to Bruni. He reacts to the stimulus and makes me smile. He doesn't let me forget he's with me."

John looked at Bruni like a proud father.

"A dog is another tool in your toolbox. He can't fix all of your problems," he said. "Having a service dog creates some other stresses, but the benefits are greater than the stresses. That look he gives me takes the stress away." He petted Bruni's big head and smiled.

"Tell me more about Bruni. How old is he?" I asked. "Can you give me an example of stress that comes from having a service dog?"

"Bruni's four and a half years old. Someone gave him up because of a divorce. He was well trained when Bill got him. Ben and Wanda fostered him, and I got to watch them train with him. The training is incredible. Bill has an eye for looking at dogs' personalities. He is amazing, and he is doing what he should be doing with Service Dogs for Veterans," he said. "One issue I have with Bruni is he can't do escalators. He stops and refuses to go because he's terrified."

John described how his life has changed with Bruni. "I am much more approachable, and my family gets a much happier husband and father. I still have issues, and there is increased responsibility with a service dog. We are together 24/7 for the most part, but the benefits are numerous. Unfortunately, you can't easily verbalize all the benefits, because it's the countless times throughout the day that you get that look from him, that instant rush of love just when you need it."

Lucas, John, Bruni, and Liam

Josh and Silo

A YOUNG MAN GOT OUT OF HIS TRUCK in the restaurant parking lot and opened the back door. A dog jumped out and sat at his feet. I knew this was Josh and Silo. I had been waiting for them. They headed for the front door side by side.

"Hi, Josh, I'm Sally," I said. "It is so nice to finally meet you."

When we had settled at our table, I said, "Silo is such a handsome boy. What kind of dog is he?"

"He is a Carolina coon dog who was abandoned at six months old with no food or water. Now he's three years old, and he's my boy!" he said. "He is such a blessing. His personality is second to none, and he's my best friend and my battle buddy. He's always there to watch my six and keep me company."

Silo settled under the table at our feet. He lay there quietly while we talked. "You enlisted in the Army when you were only seventeen. Why did you join so early?" I asked.

"I joined the U.S. Army because it was a life-long dream I had since I was a small child. I wanted to keep the family tradition of military men," he said. "I did my basic training the summer after the eleventh grade and went in after my senior year. I was in a total of ten years."

"Tell me about your time in the service," I said.

"I served in Iraq for one complete tour of twelve months, and then I did a complete tour of twelve months in Afghanistan. I was part of three special ops missions in Afghanistan with a total of five deployments," he said. "I wanted to serve in the military until I had reached full retirement, but I had severe PTSD and was hospitalized for a year. Since then I have been hospitalized three months here and two weeks there. The VA (Veterans Association) is awesome, but you have to do the footwork and stay on top of it. There is a back-up because there are so many veterans who need help."

"When did you realize a service dog might help you?" I asked.

"At first, I really didn't think I needed a service dog, until I realized that I couldn't go out and enjoy life with my family and friends," he said. "My meds only offered a little help unless I took more than needed. I would try to go out, but it ended in disaster. One time I had a minor flashback or episode, and my wife was the first one to mention the possibility of getting a service dog. I had been in counseling since 2011, and I spoke with my counselor about it. He agreed that I should get a dog. My PTSD had gotten so bad that my wife and I went through a big break up. We were separated for ten months, and that's when I decided to reach out to the VA about a dog. They put me in contact with Service Dogs for Veterans, and that's when my life changed."

"How?" I asked.

"Silo helped me and my wife get back together. She tells everybody that Silo is the reason we are back together," he said with a smile. "He has also helped me get off some of my meds, which is awesome. With Silo by my side, I can go out with family and friends knowing he is there to watch my six. I have had him now for about a year and a half, and he has saved my life. I will forever be in his debt. Silo is a blessing from God. When I first met him, we hit it off, and I teared up because it had been so long since I had felt like this."

"Josh, what tasks does he perform for you?" I asked.

"He helps me stay grounded when I have an episode or night terror. He pushes against me and gives me hugs, and that makes me focus on him and not the sights I see in my mind," he said. "He always lets me know when someone is approaching from my back, and that is something meds can't do. He never judges me when I'm hurting or need a hand. He only shows love. When I am driving, he puts his paws on my thighs, and if I get antsy, he pushes against me and calms me down."

"Silo is a remarkable dog," I said.

"Basically, he loves everyone. He has only growled at two people," he laughed. "If he doesn't like someone, then I don't either. I trust him. He's been trained to love. I've made huge strides, but I am not perfect. I still can't go to malls or Walmart."

"Does your PTSD cause the hypervigilance?" I asked.

"I hate congestion and crowds. I have been in ambushes, so when I go to a building or facility, I check it out. I look for the exits; I look at the people; I look at how I would defend it. It's overwhelming. The paranoia never goes away," he said. "PTSD is a constant roller coaster, and my wife and kids have gone through it. PTSD works on your weakest point. Everyone has a trigger, and it will break you. While overseas I saw two children killed that I had just been talking to. I held them in my arms, and there was nothing I could do. The dam just broke, and I kept seeing their faces. You start having suicidal thoughts when you can't take it anymore."

"Josh, I know the suicide rate for veterans in our country is high," I said. "I'm glad Silo is here to help you."

"Soldiers don't want to die, but we just can't get what we saw and experienced off our minds," he said. "You have been trained to be an efficient machine. It seems like they just wash their hands of you when you come out of the military. We are veterans, elder warriors, but we are still warriors at heart. Veterans have huge abandonment issues. In out-processing, it's like 'go do what you want to do,' but we are lost."

My heart breaks for Josh and all of the veterans I've interviewed. The feeling of helplessness and being lost seems universal in all of the veterans who come out with PTSD.

"Some vets fake PTSD. That pisses us off! They fake the funk. I've been out since 2016, and my mentality of life as I knew it is over. I'm glad some of the veterans who are faking the symptoms are getting prosecuted," he said.

"What advice would you give to other veterans who may be considering a service dog?" I asked.

"We have to reach out to each other and tell everything. My breaking point is different from another vet's breaking point," he explained. "What the dog does for you is so emotional. Ninety percent of the training is about obedience, and it can get boring, but Bill knows what he's doing. When you have issues with PTSD, you understand why you needed so much practice. Some veterans get frustrated when they have an episode, and the dog doesn't respond like they want. It becomes a habit after a lot of practice. Dogs are not a cure-all or miracle worker. A dog is another tool in your toolbox. The bond between vet and service dog is so strong!"

Before we ended our interview, Josh told me about his other pets, nine dogs and cats. He said, "According to Bill, Silo and I were the quickest learners in training!" Josh said he has a history of training dogs, and he enjoys helping others train their dogs.

Silo

Kim and Wrangell

I MET KIM AND WRANGELL at a local restaurant in Greer, which she suggested because it offered a special breakfast deal for veterans. On the Pups and Patriot's 5K site, I had seen a video of them, so I was on the lookout for a petite woman with a black and white dog. The door opened, and in they came. Wrangell was smaller than most of the other service dogs, but he was cute as could be. They made the perfect pair!

It was obvious from the greetings that everyone knew Kim and Wrangell. "We eat here quite often," she laughed.

"Wrangell is so cute with his black and white markings," I said. "Tell me about him?"

"I had not had a dog in thirty years when I went to adopt a dog for myself. I kept passing by him (Wrangell). I put the leash on him to walk him around, and I made sure I went out of the kennel first. I wanted to let him know I was the pack leader. He followed me out, and I knew from the beginning how smart he was. He is an American Staffordshire terrier and vizsla mix."

"Was he already named Wrangell when you adopted him?" I asked out of curiosity.

"No, I named him after the island Wrangell in Alaska where my husband, Wayne, once lived," she said. "He has a white right angle on his left side."

"He looks like a Wrangell," I laughed. "So, you adopted him first and then went through service dog training later at Service Dogs for Veterans?"

"Yes, we started in the intermediate training class, and I have had him for two years," she said. "He protects me. He knows when I am getting agitated and uncomfortable. Wrangell is a blessing from God, and we have such a bond." She looked under the table like a mother watching her child. "He knows when I'm down. He gives me a 'Paw-Paw-Kiss-High Five!'"

Our breakfast came, and we enjoyed pancakes and lots of coffee.

"Wrangell is also a trained therapy dog. We went through the therapy dog program, Paws2Care, after he finished his service dog training. When he does therapy at a local retirement home, that tail doesn't stop wagging. Everyone loves him," she said with a smile. "We also help with Patriot Hospice and the Patriot Guard Riders.

When I go to Patriot Hospice, he is there to help me more than the patient."

"What else does he do?" I asked.

"When I am in the hospital because of my health problems, he lies in the bed on top of me. The doctors and nurses say, 'Aww, he's so cute,' but if I am taken from the room for tests, he gets upset," she said. "After a procedure, he curls up at the bottom of the bed until I wake up, and then he gets back on top of me."

"It sounds like he is very protective of you," I said.

"Wrangell has learned nine tasks during his training. He's a major kisser and likes to sit in my lap," she laughed. "He has the perfect home with me and my two cats, and they are all good friends and all black and white."

Wrangell let Kim know he needed a potty break, so she took him outside for a walk. Once they came back to our table, I asked Kim to tell me about her service in the Navy.

"I was nineteen years old when I went into the U.S. Navy. My dad was in the Navy, and he wanted me to join. I was on the USS Norton Sound AVMI, a missile research ship stationed in California. I was in a year and got an honorable discharge for breach of contract for not getting my schooling," she said.

"Did your PTSD come from the time you served on the ship?" I asked.

"My father was in cryptology while in the Navy for twenty-one years, and I wanted to please him," she said. "He was abusive to me when I was a child, and that's where my PTSD comes from. He was very controlling and has caused me to have many surgeries. When I was five years old, living in Japan, I had my first ambulance ride and was in the hospital for three days. In my adult life, I've almost died six times."

"I'm so sorry to hear that you were abused," I said. I know Kim saw the shock on my face.

"God brought me out of that, but I carry lots of scars," she said. "I tried to lead my dad toward God when he was dying. On his deathbed, he weakly raised his arm, grabbed my hair, and punched me in the chin. I had a spear in my heart, but I grabbed his fist and kept kissing it and saying, 'I love you, Daddy.' I hope he was saved at the end, but I will live with that forever."

I didn't know what to say. I thought her PTSD had come from her military experience. It broke my heart to hear her sad circumstances. To change the subject, I said, "Let's talk about the activities you and Wrangell do for the community."

"We are involved with the Patriot Guard Riders. We ride on motorcycles and carry big American flags to funerals for fallen veterans and officers. Wrangell was upset at first when a twenty-one-gun salute was given, but he is doing much better now. All branches of the service and all kinds of veterans come to these funerals, even Hell's Angels."

Kim told me she had a Harley and sold it recently. I couldn't imagine tiny Kim on a big Harley, but she has a huge heart for helping others, so I could see her taking part in the funerals.

"There are times when I just don't want to be here anymore. Without my dog and cats, I don't think I would be here now," she said. "Wrangell helps me get my focus off myself."

"So, you and Wrangell met Bill and started in the service dog training program, right?" I asked.

"Bill's organization is wonderful!" she said. "I loved the idea that we trained alongside our dogs. It has been a true blessing getting Wrangell, meeting Bill, and going through training."

"Was Wrangell easy to train?" I asked. "He seems so calm and well trained now."

"Wrangell is full of energy, and he is excitable, but he is very smart and attentive. When I don't feel like taking him for a walk, I put him on the treadmill. I just say 'treadmill,' and he climbs on.

I set the timer for fifteen minutes, and he is all happy and wiggly-butt," she said. "One time he got tired and his foot slipped. It scared him."

I can tell Wrangell is all business when he wears his vest. "Kim, you and Wrangell are perfect together. I can see the strong bond between you."

"He has given me a quality of life that I didn't have before. If I am down, he pulls me out of it," she said. "We are together twenty-four hours a day. He's even in the church choir and Bible study with me! I go to the same church as Bill and Pam, and we've all become good friends."

Wrangell

Laura, Paul, and Hope

Paul and Hope

I GOT TO THE CLEMSON ARMORY before Paul and Hope arrived for our meeting. He was a young man with longish hair and a big smile on his face. Hope, a yellow Lab, was right beside him. Our first meeting had been rescheduled because Hope had injured herself in an accident while playing outside.

We decided to meet in a back room where we could talk. We both got settled, and Hope sat at Paul's feet. "Hope looks like her chest wound has healed," I commented.

"Yes, she is coming along fine. You can barely see where she had stitches," he said. "Where do you want to start?"

"Let's start with your military background," I said.

"I was nineteen when I joined the Marines in 1998," he said. "I chose the Marines because I knew it would be hard, and I would get to go all over the world and do cool things. I was stationed at Camp Lejeune."

"How long were you in the Marines?" I asked.

"I was on active duty for four years, and when my enlistment was up, I joined the National Guard in Seneca, South Carolina, which is where I'm from," he said. "At first, I worked with radars. After that, I was a full-time recruiter, and then a drill sergeant. I retired as a supply sergeant with a total of nineteen years."

"You were in the service for a long time!" I said. "Did you retire for medical reasons?"

"Yes. I was medically retired as an E-7 Sergeant First Class. My story is a little different from the other veterans. In 2015, I was diagnosed with multiple sclerosis. I was having a hard time getting around, so they gave me a foot lift to keep me from tripping. It kept getting worse, so next I got a walker. I guess it looked bad for a soldier to be on a walker, so they started the process to put me out of the service."

"Tell me how you got your service dog, Hope," I said.

"One morning my wife got a call from family services, and they said a Lab breeder had donated a dog for a soldier. They asked if I wanted a service dog. My wife and I both said, 'Yes!' The next day they brought me a cute, untrained puppy. I named her Hope. In other words, I hope she doesn't let me fall!" he laughed. "After bringing us the puppy, the lady from family services got sick and quit, so I was stuck with an untrained puppy."

"How did you learn about Service Dogs for Veterans?" I asked.

"The VA gave me Bill's number, and he saved us. They helped me train her to help me get around physically and mentally. Hope was only six months old when we started," he said. "We had to do some different training so she could help me with mobility. Usually you have five outings during training, but Hope and I had nine outings. We graduated about the time the class behind me finished the program."

"So, were you a Marine during 9/11?" I asked. "What was that like?"

"I was in Kosovo when 9/11 happened. Everything was locked down when we heard it on the radio. It was kind of unreal. I immediately called home, and my little brother answered the phone. I found out later he didn't remember to tell my mom I had called. She was mad."

"Why did you decide to join the National Guard?" I asked.

"I missed being in the service with the guys. I enjoyed my time, and that's why I joined the National Guard. I did it because I liked it. It was a fun nineteen years!" he laughed.

"No wonder you made a great recruiter! Your smile is infectious," I said.

"If you had seen me two years ago, you would have thought 'What's wrong with that guy,' because I could barely get out of bed. Then I was diagnosed with Crohn's Disease, and I had to stay in the hospital," he said. "My wife said I looked like I was on a concentration camp diet, I was so thin. Once I got better, she took me on a cruise to fatten me up."

"Did you take Hope with you on the cruise?" I asked.

"No. I left her at home with the kids. I'll have to tell you what happened while we were gone. You may have seen it on the news," he said. "My son had Hope in the car with him when he had a wreck. She somehow got out of the car and took off, and he was taken to the hospital. My wife has this app on her phone that showed our son was in the hospital, so we called home immediately. Everyone was looking for Hope. It was on the news that a veteran's service dog was missing. The police and fire department were looking for her for hours during twenty-degree weather. Hope was spotted by the police, and she finally jumped in their car."

"I bet you were scared!" I exclaimed.

"My son was not seriously hurt, and we were upset when Hope was missing. It took four hours to find her and get her in the car. She has always been such a good, calm dog. I couldn't have handled a crazy dog. Even as a puppy, she was calm. I never really understood about dogs before Hope, but she is a part of me now."

After our interview, I got this email from Paul that brought tears to my eyes.

Thank you for sitting down and talking with us today. I hope we helped. Thank you for what you are doing by telling the stories of the vets. If you can work it into the book, please tell the readers I said thanks for my opportunity to serve, and thank you (the country) for taking care of me and my family so well, since I can't do it anymore.

Paul and Hope

Steve and Barrett

I ANXIOUSLY WAITED FOR STEVE AND BARRETT at Panera Bread in Greenville. Steve was my last scheduled interview, and he was a Vietnam vet, so he was older than the other veterans I had interviewed. I sat by the window, looking for an older man with a dog. A large pickup truck pulled up, and I thought I saw a dog's profile. That must be Steve, I thought to myself, but he wasn't getting out of the truck. Finally, the door opened, and a tall, thin, gray-haired man got out. He opened the back door, and a large black Lab jumped from the back seat. Barrett immediately sat in front of his master and looked into his face as if saying, "What do you want me to do?"

Steve had such a serious look on his face it made me nervous. I said hello, and he settled Barrett under the table while I took care of our order.

When I seated myself across from him, I said, "I love the name Barrett. It is such a distinguished name. It sounds like an English butler's name."

He finally smiled. "I love his name, too, and it fits him."

"It certainly does," I agreed. "He looks so noble. Tell me about Barrett. How did you get him?"

"Actually, I didn't want a dog. I didn't need a dog," he said. "I met Bill, the founder of Service Dogs for Veterans, at a Vets Helping Vets program, and he brought dogs with him. I got up with Bill and asked if I could come to training sometime. I was working with a friend who had a miniature Doberman he wanted to train. Bill said that would be great."

"One night at training, Bill had Barrett with him. He was between us, and Barrett kept staring at me. He picked me! To make a long story short, at the time I met Barrett, I was living where there were some other dogs, but we worked it out. I've had him almost a year now. I finally adopted him. He's something!" he said, laughing. "He gets me into spots that I wouldn't have been in, and they are good spots."

"What do you mean?" I asked.

He leaned back in the booth. "For example, Barrett got me introduced to Charlie Daniels!"

"What! I've got to hear this story!" I said.

"I was at Upstate Salute to Warriors where we had a Service Dogs for Veterans booth set up. I had Barrett with me with all of his stuff - toys, bed, etc. Bill said I needed to get out and mingle, so I did," he said. "I met a girl in a wheelchair with seven fiddles stacked in her lap. 'I can't let you carry all these fiddles,' I said to the girl. 'I'll carry the fiddles, and Barrett and I will follow you.' I followed her to a tent. It was some kind of ministry out of Spartanburg, South Carolina. The next thing I know, here comes this guy with a big cowboy hat on. It was Charlie Daniels! We started talking about dogs, his farm, and vets. 'You're coming with us!' Charlie said. We went up on the stage, and he handed us a fiddle in front of the audience. He autographed it - To Steve and Barrett from Charlie Daniels. Barrett was such a show-off that day! We met a lot of people."

He smiled and rubbed Barrett's head.

"Barrett loves kids. I have a five-year-old grandson, and they love each other. I let him train Barrett like I do," he said. The serious demeanor was disappearing. I could actually see him glow with pride as he talked about his grandson.

"What other spots has Barrett gotten you into?" I asked.

"I live in Anderson, South Carolina, where I'm friends with a vet who has a rat terrier named Pepe. We met at Vets Helping Vets, and we started training together. He's a seventy-two-year-old Vietnam vet who takes care of his ninety-two-year-old mother. We work on training about three times a week. Vets Helping Vets is a group of friends I never expected to have. With Barrett by my side, I talk to people all the time. Actually, I did need a dog! My social worker and psychiatrist said Barrett was probably the best thing I could have done."

"Steve, you have such a wonderful story to tell," I said. "How does Barrett help you?"

"Barrett knows when I'm down. He places his head in my lap. He chills me out when I'm anxious or frustrated. He works all the time."

"How does he do that?" I asked.

"Barrett hasn't cured anything, but he's made it easier to get through the day. I don't have many bad days anymore. He doesn't let me out of his sight. He fills my heart with love," he said. "He helps train other dogs, and he helps other vets. I wish my life were different. Some stuff in the service really messed me up. There was a time when I almost wasn't here. I ruined a lot of my life living in the past."

"Has Barrett helped you live in the future?" I asked.

"I shouldn't be here in all reality. Now every day is worth doing something good. I have gifts: Barrett, my grandson, and friends," he said. "I've spent most of the last four years taking care of my grandson. I have a thirty-two-year-old son that I haven't seen in six years. I have had a lot of loss in my life, and I've lost a lot of good friends, but now I have new friends."

"Steve, I'm glad things are working out for you now and that you have Barrett to help you navigate through life."

"God gave me a gift. He shows me what to do. Barrett and I get to watch the sun come up every day," he said.

"A day came along when I knew I needed to do something. I have found my niche --what I am supposed to do. Here it is!"

I leaned in. "What are you going to do?" I asked.

"I have a good friend in Vets Helping Vets who needs a kidney, and I've been getting tested to donate my kidney. Lots of testing! This part of my life is about giving back. I talked to a living donor nurse on Friday, and my friend's case went to the board. I should know something by the end of the day today," he said, sitting a little straighter. "I have a lot of friends now; 100 guys who will be there for me. Two guys offered to take me to Duke. My friend Gary and his dog Pepe offered to take care of Barrett while I'm in the hospital. I only trust a few people with Barrett, and Gary is one of them."

Steve looked me in the eye.

"It's a God thing!" he said. "I did a lot of praying about it. I'm ready, and I'm not afraid. I've always wondered what I was supposed to do with my life."

"Steve, I know your dear friend is blessed to have you in his life." I felt like I could burst into tears as I sat in front of this unselfish man.

"Let's talk about your time in the military," I said, and I saw him tense up.

"I was seventeen when I enlisted. My best friend had signed up with the Marines, but he never came back," he sighed. "I joined the U.S. Navy and had a six-and-a-half-year enlistment. I was a Boiler Tech, and on a ship for three-and-a-half years. After the Navy, I went over eight years in the Army National Guard. I worked around ammunition and demolition, and I burned out. I was getting drunk all the time. When I joined the Navy at seventeen as a Boiler Tech, I was at the bottom of the ship where I got attacked!"

I got the feeling he didn't mean to tell me about the attack.

"I can't believe I said that!" he admitted. "It put a shell around me. I hit the bottom in 1989 when I was living in North Carolina. I was ready to end my life," he said sadly. "I called this guy I knew in the middle of the night to tell him my plan. I didn't expect him to answer the phone, but he did answer, and he saved my life! It was a God moment."

"I'm glad he answered the phone, too. Your story makes me so sad, but I'm glad everything has worked out," I said, not knowing what to say. I had just met this man an hour ago, and he was telling me the intimate details of his life. I'm sure he could read my distress at hearing his sad story. "Let's get back to how you met Bill."

"A friend of mine was a sniper, and he was having a tough time. He finally gave up alcohol. He died sober in a motorcycle accident. I got to know his wife, who introduced me to Bill, who's a saint."

"Why do you call him a saint?" I asked, already knowing the answer.

"Bill wants to give back, like me. There's a need. When I got out of the military, no one knew there were twenty vets a day committing suicide," he said.

"It breaks my heart that we are losing so many of our veterans," I said. "This is why I am writing this book. There is hope, and it's often found in a four-legged friend who's there to help you heal."

"When I wake up in the morning, Barrett's right there. He's there when I'm up and there when I'm down. Service dogs aren't a cure, but they make life a lot easier. He is protective of me, but I'm more protective of him," he said, petting Barrett's head. "I was telling my grandson, who's going into first grade, that DOG is GOD spelled backwards."

"Steve, what are your plans for the future?" I asked.

"Bill and I would like to see Vets Helping Vets all over the state, and I plan to continue to go to Service Dogs for Veterans events. I want Bill to be successful. Last year we went to Puppies and Patriots 5K run in Clemson. It was great, and all the money went to Service Dogs for Veterans. It's held Memorial Day weekend."

"You mentioned that Barrett had some service dog training before you got him," I said.

"Barrett was one of Bill's first dogs, and he had been trained for an autistic child, so he was trained when I got him. I had to retrain him as my service dog, but he was easy to train. The problem is he has too many friends. People friends! The girls come running when they see Barrett. He's a gift, and I've been blessed. I can show other vets that this works. This is the key!"

"He is a wonderful dog. I love his vest," I said.

"I'm having a vest redone for him. I'm going to add a flag and a patch. I got this vest when we finished our training. A friend of mine has a sister who's a seamstress, and she's fixing a vest that doesn't rub his tummy."

"He's definitely a good-looking boy!" I said.

"I live in pain. My hands hurt daily, but I hardly know it when Barrett's around. We're here to help!" he said with a big smile.

Endnote: At the end of the day Steve and I met for the interview, he got the approval to donate his kidney to his friend. He said, "On October 3, 2019, I went into a Marine! Now we are 'brothers from another mother.'"

Steve and Barrett

Tim and Marla

I MET TIM AND MARLA at the Landrum Public Library. His ten-year-old daughter, Allison, came with him. We got a private room where we could talk, and she explored the library. Tim served in the Army as an assistant chaplain for fourteen years, so I wanted to hear his story.

"Why did you decide to join the military?" I asked, looking down at Marla lying at his feet. She is solid black with beautiful, pensive dark eyes.

"Interesting story here! I never wanted to join the military," he said with a laugh. "In high school I had begun to feel led to the ministry, so I told all the recruiters who called that the only way I would enter would be as a chaplain or chaplain assistant. This was quite effective in stopping all calls from all branches; however, there was a local recruiter, I later discovered, who took that as a challenge. Almost a year after I graduated from high school, I received a call from this man. He stated he had a chaplain assistant position with a unit fairly close to where I was living. I was floored, and asked if I could think about it. I spoke with my parents and my priest, and we all prayed about it. I went to meet him, and it turned out he was a devout Christian who had seen my request and taken it to heart. We hit it off, and after meeting him and much prayer, we decided it was in God's hands."

"Did you begin as a chaplain?" I asked.

"I had had major reconstructive chest surgery four years prior, and I wasn't even sure I could get into the Army. I took the ASVAB test and scored well. The recruiter thought the only issue would be that I was underweight. We signed the form, and I ate lots of bananas and drank milkshakes for a week before going to Fort Jackson's Military Entrance Processing Station. It took five more trips over a month and a half before I was accepted into the military. It took another two years to become a chaplain assistant."

"Where did you serve?" I asked.

"I was in the U.S. Army Reserves primarily stationed in Spartanburg, South Carolina," he said. "In 2005, I was mobilized under Operation Enduring Freedom and spent four years at the Conus Replacement Center at Fort Benning, Georgia. This unit deployed 500 individual augmentees, members of the military assigned to special duties, and redeployed 250 individuals weekly. We were the last faces they saw on U.S. soil before they headed to over thirty-six different areas around the world. Primarily, they went to Iraq and Afghanistan. As the chaplain assistant, it was my duty to speak to each person and make a determination as to whether they were spiritually, emotionally, and mentally ready to either go to war or return to civilian life."

"What caused your PTSD?" I asked.

"My PTSD comes from Fort Benning when I was counseling as a chaplain assistant. I was on call 24/7. I did most of the counseling, and they called me 'Chaplain Tim.' It is called compassion fatigue. You are so empathetic that it takes over your body," he said. "We also had a young vet killed in a terrible wreck which deeply affected me. When I got out, I felt there was no purpose anymore. I was diagnosed with PTSD by proxy."

"I didn't know that you can get PTSD by counseling others going through difficult times," I said. "How did you know you needed a service dog?"

"Originally, I had a wonderful dog named Bear, who helped me through a rough period. I wanted to get him trained as a service dog, but because of his age, we determined it wasn't feasible," he said. "Unfortunately, he was hit by a car and died on the way to the emergency veterinary clinic. By the end of the day, I was joining Service Dogs for Veterans to find a new service dog."

"Marla is so calm and sweet," I said, gazing into her beautiful eyes as she looked up at me.

"Marla had been through training with another veteran, and her name was Darla before I got her. The veteran dropped out of the program, and I needed a dog, and she needed a veteran," he laughed.

Tim refers to Marla as his American Express Card, so I asked him why he calls her that.

"I never leave home without her!" he exclaimed. "She is the most wonderful dog. She helps with my back issues by retrieving items from the floor; she braces when I need help to stand, and I have a mobility harness when I need extra support walking. For my PTSD, she will reflect my level of emotionality to tell me how I currently am. If I get highly stressed or agitated, her behavior mirrors it to allow me to check myself and ensure that I am at an appropriate emotional level. She also pushes against me and makes me give her affection to help lower those levels."

"Tim, has your life changed for the better with Marla by your side?" I asked.

"I like to tell the story of when my eight-year-old daughter and I were in a Walmart, and a random person asked how Marla helps me. My darling little girl looked at her and said, 'My daddy doesn't get mad at me anymore.' With tears in my eyes, I walked away, holding my daughter's hand."

Tim told me he had been through some rough times over the years. He and his wife got a divorce in 2015 after eighteen years of marriage, and he moved in with his parents. His mother passed away six months later. His son went to live with his ex-wife, and his dog Bear was hit by a car. He obviously had a lot of stress in his life, along with PTSD.

"Tim, I know you are going to Converse College in Spartanburg, South Carolina, and working on a degree," I said. "Tell me about your plans for the future."

"I will graduate from Converse next May with a master's in Marriage and Family Therapy. Marla goes to my classes with me, and everyone loves her," he smiled. "The other students in my classes say 'Tim, I need some Marla time!' Marla is also a certified therapy dog, and she loves helping people. When I get my certification, she will be able to help me in my practice."

"You help Bill with the Service Dogs for Veterans website, right?" I asked.

"When I first met Bill, he was extremely overworked, but he didn't want to let go of anything. I started attending events and talked to him about where he wanted the organization to go. My major was marketing, so I did a two-page marketing plan for him," he said. "Bill and I recruited about seven or eight women to be our marketing team. We finally got the website up. One of our largest donors said our website didn't look like a site where he would make a large donation. Bill said, 'Fix it!' We did and got the donation. One of the first things people see is the website."

Tim's daughter came into the room after exploring the library and laid on the floor next to Marla. Marla gave Allison a kiss, and like a stern father, Tim reminded Marla she was still working. There is a strong bond between Marla, Tim, and Allison.

"I am the networker, and I get out the word to the veteran communities. With school, I have had to pull back a little bit. Bill and I go to all events, and at the current time, we have had 295 applications since we opened," he said.

"I have spent a lot of time looking at the pictures and reading testimonials on the website," I said. "Tell me one of your favorite stories about Marla?"

"It was the weekend of the Animal Fair in Landrum, South Carolina. This is one of the big fundraisers for us and the Foothills Humane Society. I had already bonded with Marla just from being around Bill. I met Bill at the event, ate a wonderful meal, and met Lennie Rizzo, who wrote an article about the organization. Bill got up on stage to take Marla through her maneuvers. He had her in a sit/stay," he said. "Bill said, 'Marla will stay until I release her.' I was taking pictures from the audience when a lady asked if Marla was my dog. She said, 'She watches you wherever you go.' I responded, 'She goes home with me today!' When Bill released her, she jumped off the stage and ran to me. Bill was embarrassed, but I said, 'We couldn't have planned it any better. This shows the bond between us!'"

"She obviously loves you very much!" I laughed.

"She loves Bill, too, but now she's my dog!" he said with a grin on his face. "When I'm down, she gives me a kiss and calms me down. Her energy reflects mine and makes me check myself emotionally."

"Tim, I am so glad you have Marla by your side, and you will make a wonderful counselor when you finish your degree," I said.

He smiled that big smile. "Marla and I are happiest when we are helping someone!"

Marla

Service Dogs for Veterans

by
Mary-Ellen Gregory (MEG)
Community Outreach Coordinator

SERVICE DOGS FOR VETERANS (SD4V), founded in 2014 by Bill Brightman, is a nonprofit organization providing a life-changing service for military veterans in Upstate South Carolina and Western North Carolina. The mission statement is "To complement the traditional treatment of veterans with PTSD, TBI, and MST by training them and a dog to become highly effective ADA (Americans with Disabilities Act) compliant service dog teams."

SD4V's purpose is to address an unmet local need for providing service dogs to disabled veterans with PTSD (Post Traumatic Stress Disorder), TBI (Traumatic Brain Injury), and/or MST (Military Sexual Trauma). Partnering with Dog Trainers Workshop in Fountain Inn, South Carolina, a program was designed that trains the veteran to train his or her dog, creating an inseparable, bonded team. The website is www.sd4v.org.

In 2017, Service Animal Project (SAP), a Tryon, North Carolina, based organization that evaluated, rescued, and donated shelter dogs to a similar program in Florida, was looking for a local service dog program to work with. SAP merged with SD4V creating a win-win for both organizations.

As of the end of March 2020, SD4V has graduated sixty-three military veterans from the program, allowing them to regain normalcy in their lives with the help of a service dog. These men and women experience a transition from extreme depression, isolation, anger, and thoughts of suicide to a new life of joy, anticipation, and return to family involvement. With a trained service dog at their side, they are able to mitigate the limitations caused by PTSD, TBI, and MST.

Targeted Population

SD4V serves military veterans within a two-hour drive of Greenville, South Carolina, who are willing to commit to a seven- to eight-month program that requires attendance at weekly training sessions. Veterans from any wartime era who have been diagnosed between fifty percent and one hundred percent disabled by the Veterans Administration (VA) are eligible to apply. Typically, SD4V's clients have been treated by the VA for years, but have not seen the improvements needed to live a more normal life. By training the veteran and dog to become a well-bonded service dog team, the veteran's debilitating wartime symptoms can be alleviated.

Training Program Strategy and Goals

SD4V's program prepares the veteran to be a trainer and handler of his or her dog. The three-phase process over the period of seven to eight months transforms the veteran and his or her dog into a service dog team which meets the compliance guidelines of the ADA.

After the veteran's initial contact with SD4V at www.sd4v.org/veteran-info, the process begins.

Phase I consists of intake and processing over a period of one to two weeks. This requires the veteran to complete an online application and submit appropriate VA-issued documentation related to his or her disabilities. Home visits are made to review the veteran's needs and lifestyle. SD4V acquires a fully vetted shelter dog that gets evaluated as to suitability for the program. Should the veteran choose his or her own dog for training, then suitability assessment is made with that dog before it can begin training.

In order to provide a baseline for the veteran's symptoms, he or she is asked to complete a PCL-5 questionnaire. This PTSD checklist is a twenty-item self-report measure that assesses the presence and severity of PTSD symptoms by identifying a numerical value for each of the questions. The PCL-5 can be used to quantify and monitor symptoms over time to screen individuals for PTSD. The values that are recorded on the PCL-5 document are totaled to give a baseline value to their symptoms at the time they enter the program.

Phase II is the companion dog program phase and covers three to four months. The veteran and dog attend weekly beginner and intermediate dog obedience classes. The veteran is required to practice skills learned in each class and document and submit training time each week.

At the end of Phase II, assessment is made to determine suitability to move into the service dog program phase. If it is determined the veteran does not move to phase III, he or she graduates the program as a companion dog team. These dogs are well behaved and provide comfort and a sense of security at home and in dog-friendly public settings. They do not, however, have public access rights under the ADA, and the dog does not receive a service dog vest.

Phase III takes five months and begins with awarding the dog a 'Service Dog In-Training' vest to wear at all training times. The veteran and dog team start with six weeks of advanced obedience training, then six weeks of task training (concurrent with advanced classes) and finally eight weeks of public access practices.

During the public access training, the handler and dog apply all they've learned while in a variety of public settings and under the supervision of a professional trainer. They must demonstrate they work as a team, the dog knowing what to expect from the handler, and the handler knowing what to expect from the dog. To comply with the ADA, the dog must perform learned tasks either by command or by its own initiation. They must do this repeatedly, despite distractions found when in the public domain.

Upon successful completion of the public access phase, the team graduates. The 'In Training' patches on the vest are replaced with 'Service Dog' patches, and a graduation certificate and an identification card is awarded. A post-graduation PCL-5 questionnaire is completed, and a post-graduation assessment interview is conducted with the veteran to provide feedback regarding the program.

Other Programs Available to Graduates

SD4V provides three additional programs that veterans can take advantage of upon graduation.

Workforce Development Program: Graduate veterans often find themselves becoming more social and outgoing with their service dogs by their sides. Sometimes the team is ready to look for employment or enroll in education programs. SD4V's volunteer human resources professional will assist the veterans with job searches, resume writing, and mock interviews. They will also assist the veterans by helping to educate area employers about ADA and the realities of accommodating ADA-compliant service dog teams.

Additionally, thanks to a grant received, SD4V can offer graduates an opportunity to experience an aptitude assessment with the renowned Johnson O'Connor Research Institute in Atlanta, Georgia. Armed with new insights about their aptitudes and interests, these veterans can move forward with their lives with greater confidence.

Regional Monthly Meetups: During the Phase III training, the veterans forge a bond with other veterans in the program and rediscover the military camaraderie they enjoyed during their active duty. SD4V has instituted monthly regional meetups in several upstate counties where graduates can come together. These meetups are organized and led by veterans who have graduated from the program and provide an opportunity for those attending to continue the bonds they made during training classes.

Therapy Pet Program: Through SD4V's therapy dog partner, Paws2Care (https://paws2care.net/), veterans can have their service dog certified as a registered therapy dog that can be taken into hospitals, hospice facilities, retirement homes, behavior centers, and schools. The veteran shares the companionship of his or her therapy dog with people who can no longer enjoy the company of their own pets. By petting and talking to the therapy dog, these people can be distracted from their own situation, ultimately providing an emotionally positive result. An SD4V graduate functions as the facilitator for this program.

Measurable Impacts

Measurement of success is established by calculating the improvement in symptoms as recorded by the veterans during their PCL-5 Self-assessments before and after training. Historically, veterans have experienced, on average, a forty-five percent reduction in war-related symptoms upon completion of the program. Many times, the veteran will experience a similar reduction in his or her prescription medications and therapy sessions.

During the post-training assessment, veterans report that being part of a service dog team results in significant mitigation of the life-limiting effects of their service-related symptoms. Many of the veterans describe the result as "life-changing," including:

o   Reduction in thoughts of suicide
o   Reduction in need for prescription medications
o   Reduction in alcohol consumption
o   Discovery of a career path or desire to pursue higher education
o   Improvements in their marriage and family relationships
o   Ability to attend and enjoy public events with no panic attacks

Cost

SD4V is blessed to have the support of numerous local and national foundations, businesses, and individual donors. These donors allow SD4V to provide ninety-seven percent training scholarships to veterans entering the program. The out-of-pocket cost for the eight-month program is $200. The first $100 is paid prior to starting the beginning obedience class, and the second $100 is paid when the team starts the service dog training portion of the program.

Thanks to the countless hours contributed to the organization by eighteen dedicated volunteers, SD4V can apply ninety-five percent of funding received directly toward program costs.

Service Dogs for Veterans (SD4V)
P O Box 965
Taylors, SC 29687
(864)230-3800

Acknowledgments

WRITING *HEROES OF THE HEART* has been a blessing from God and privilege for me. While interviewing the veterans and their service dogs, I got a glimpse into the lives of real-life heroes. I can't thank you enough for sharing your stories, happy and painful, with me. Thank you, Kim, Ashley, John, Paul, Tim, Josh, and Steve, for your service, for your courage, for your resilience, and for your honesty. I hope you know how much I appreciate our time together as you answered my many questions. Also, thank you for the wonderful pictures you sent for the book.

Thank you to my new canine friends that are heroes and healers. Bruni, Wrangell, Marla, Silo, Barrett, Aubrey, and Hope, you are all "Top Dogs" in my eyes! Thank you for sharing your love, kisses, wiggles, and wags to comfort our veterans.

Bill Brightman, a good friend and an outstanding human being, saw a need to help veterans when he moved to Greenville, South Carolina. He founded SD4V and has worked tirelessly with our veterans and rescue dogs since 2014. People like you make this world a better place for everyone. Thank you for answering my many questions, for meeting with me, and for inviting me to watch training. I hope and pray *Heroes of the Heart* will encourage more veterans who are suffering with PTSD, TBI, and MST to seek you out.

To my old friends, Linda Williams and Ann Goodheart, thank you once again for your help and encouragement. It was only fitting that you two should write the Foreword to *Heroes*. You are both champions of our veterans and rescue dogs.

Thank you Mary-Ellen Gregory (MEG) for writing the chapter about the organization. As Community Outreach Coordinator for SD4V, you know the workings of the organization inside and out. I also want to thank Sharon Rose, a volunteer at SD4V, for helping you with the chapter.

Thank you once again, Gayle Wafrock, for working your photo editing magic on the photographs. You are a wizard when it comes to photography, and I can't thank you enough for your help, dear friend.

Another indispensable person is my editor, Carol Crawford. Thank you for your thoughtful suggestions and excellent editing skills. I couldn't have done it without you.

Thank you to Tim Davis, my old friend, who is an excellent illustrator, writer, and designer. I love the cover you created for *Heroes*.

I have to thank my good friend and technical adviser who has helped me with all three books. Col. Tom Davis of Old Mountain Press has been a lifesaver since I met him at a conference in 2017. I don't know if there ever would have been a first book if it weren't for you. Thank you for your patience and friendship.

I can't forget about my family's support. My husband, Hugh, has been my biggest cheerleader since I started writing. He always encourages me and offers good advice. A veteran himself, he especially loves *Heroes of the Heart*. My two rescue dogs, Rudy and Einstein, cheer me on as well. They love books about dogs!

I want to thank all of the volunteers, donors, trainers, veterinarians, rescuers, and organizations who support and believe in the good work SD4V does in the community. Without your support, nothing would be possible.

Finally, and with heartfelt love and gratitude, thank you to all our special canine friends who love us unconditionally and make our lives better every day. You are our heroes! Remember, DOG is GOD spelled backwards! Sally

(Photo by Gayle Wafrock)

About the Author

SALLY HURSEY'S LOVE OF DOGS and veterans began when she was a child growing up in a small southern town surrounded by lots of pets. Her father, who served in the U.S. Army in Northern Africa, Sicily, and Italy during WWII, instilled a love of the military and service to America in his four children. While Sally was growing up, her father continued to serve in the South Carolina Army National Guard. Her love of veterans and dogs is evident in her third book, *Heroes of the Heart*.

Her first book *Molly to the Rescue*, is a children's book about a local rescue dog that becomes an elite service dog for a veteran suffering with PTSD. Her second book, *Big Sid,* is a picture book about sibling rivalry, dog style!

Sally holds a BA in English from the University of Georgia, an M.Ed. from Converse College, and an MLIS from the University of South Carolina. She was formerly a high school English teacher and a high school library media specialist. She lives in the Carolina mountains with her husband, Hugh, and two rescue dogs, Rudy and Einstein. Visit Sally at www.sallyhursey.com or on Facebook-Sally Hursey-Author.